SCIENCE TOPICS

Chemicals in Action

ANN FULLICK

SKAHA LAKE MIDDLE SCHOOL
LIBRARY
110 GREEN AVENUE WEST
PENTICTON, B.C. V2A 3T1

Heinemann Library
Des Plaines, Illinois

© 2000 Reed Educational & Professional Publishing
Published by Heinemann Library,
an imprint of Reed Educational & Professional Publishing,
1350 East Touhy Avenue, Suite 240 West
Des Plaines, IL 60018

Customer Service 888-454-2279

All rights reserved. No part of this publication may be reproduced or transmitted in any form or by any means, electronic or mechanical, including photocopying, recording, taping, or any information storage and retrieval system, without permission in writing from the publisher.

Designed by AMR
Illustrations by Art Construction and Phillip Burrows
Printed in Hong Kong

04 03 02 01 00
10 9 8 7 6 5 4 3 2 1

Library of Congress Cataloging-in-Publication Data

Fullick, Ann, 1956-
   Chemicals in action / Ann Fullick
      p. cm. – (Science topics)
   Includes bibliographical references and index.
   ISBN 1-57572-776-5 (library binding)
   1. Chemistry Juvenile literature. I. Title. II. Series.
   QD35.F85 1999
   540—dc21   99-27944
                                                           CIP

Acknowledgments
The Publishers would like to thank the following for permission to reproduce photographs:
J. Peter Gould, pp. 5 (all), 7 bottom, 8; Allan Cash, pp. 6, 9 top; Science Photo Library/Charles D. Winter, p. 7 top; Science Photo Library/John Durham, p. 9 bottom; Science Photo Library/John Mead, p. 11 top; Jo Malivoire, p. 11 bottom; Environmental Images/Robert Brook, p. 12; Still Pictures/Hubert Klein, p. 13; Science Photo Library/Volker Steger/Peter Arnold Inc., p. 14; Science Photo Library/Andrew Syred, p. 17; Still Pictures/Andre Maslennikov, p. 19 left; Still Pictures/Ray Pfortner, p. 19 right; Still Pictures/J.P. Sylvestre, p. 20; Science Photo Library/ESA, p. 23; Ancient Art and Architecture, p. 24; Still Pictures/Dylan Garcia, p. 25; Still Pictures/Paul Gipe, p. 27; Mary Evans Picture Library, pp. 28, 29.

Cover photograph reproduced with permission of Science Photo Library/John Mead.

Our thanks to Kris Stutchbury for her comments in the preparation of this book.

Every effort has been made to contact copyright holders of any material reproduced in this book. Any omissions will be rectified in subsequent printings if notice is given to the Publisher.

Any words appearing in the text in bold, **like this**, are explained in the Glossary.

Contents

Chemical Reactions 4

The Reactivity Series 6

Pulling Rank 8

Using Reactivity 10

Oxidation and Reduction 12

Acids and Alkalis 14

Making Salts 16

Rain and Rocks 18

Water 20

Pollution 22

Materials 24

Fossil Fuels 26

Looking Back 28

Glossary *30*

More Books to Read *31*

Index *32*

Chemical Reactions

Everything on Earth is made up of **atoms**. Sometimes the atoms are found on their own, but often they are joined together to form **molecules**.

Chemical combinations

Atoms join together to form molecules. A molecule may contain atoms of the same kind or atoms that are different. If all the atoms are the same, the molecule is an **element**. If the atoms are different, then the molecule is a **compound**. Each chemical compound has its own **formula**, and this tells us which atoms make up the compound and how many of them are combined in each molecule. For example, O_2 is a molecule of the element oxygen and contains two atoms of oxygen. H_2O is a molecule of the compound water and contains two hydrogen atoms and one oxygen atom.

Joining together

How do atoms join together and why do they join in particular combinations? To understand this, we need to use **models** to help us imagine what is going on inside the atoms themselves. To understand the way atoms join up, imagine each type of atom having a certain number of "arms" available to join on to other atoms. The number of arms depends on how many **electrons** the atom has. For a **stable** molecule to be formed, all of the arms of all of the atoms involved must be joined to others.

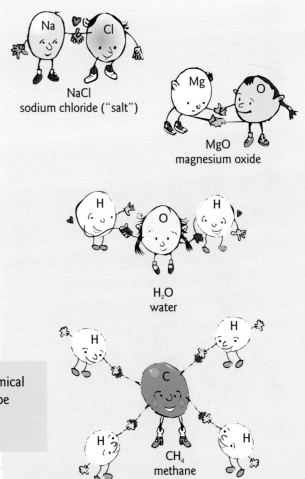

NaCl
sodium chloride ("salt")

MgO
magnesium oxide

H_2O
water

CH_4
methane

▶ The "arms" on these model atoms link to form chemical bonds. To make a compound, all of the arms must be used up to make the right number of bonds.

Stay in balance!

Before and after any chemical reaction, the number of atoms of each substance on one side of the **chemical equation** must be exactly the same as the number of atoms on the other side of the equation. As a result, the mass of the substances that are being reacted together is exactly the same as the mass of the substances that are formed as products of the reaction. Mass is never lost or gained as a result of a chemical reaction.

SCIENCE ESSENTIALS

Molecules contain two or more atoms joined together. Compounds are substances formed when the atoms of two or more elements combine. In a chemical reaction, the total number of atoms involved stays the same. They are just rearranged into new compounds. The mass of the **products** formed in a reaction is always equal to the mass of the **reactants**.

the mass of the product (iron sulfide) = 8.800 grams – exactly the sum of the mass of the two reactants

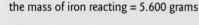

the mass of sulfur reacting = 3.200 grams

the mass of iron reacting = 5.600 grams

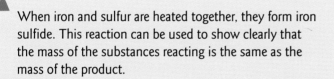

When iron and sulfur are heated together, they form iron sulfide. This reaction can be used to show clearly that the mass of the substances reacting is the same as the mass of the product.

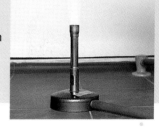

The reaction takes place when iron and sulfur are heated.

Fast or slow reactor?

Some chemical elements react very easily and quickly while others do not react with other substances at all. How easily an element reacts depends on the number and arrangement of the electrons whizzing around the nuclei of its atoms. Some arrangements of electrons are very stable and the elements tend not to react. These are called **unreactive** elements.

If an atom has a much less stable arrangement of electrons, it will react readily with other atoms to gain, lose, or share electrons and become stable. Elements like these are called **reactive** elements.

The Reactivity Series

Some metals react more easily than others, and some react slowly or not at all. By comparing the ways in which different metals react, we can predict what will happen in a whole range of reactions.

Reacting with oxygen

Some metals, such as sodium and potassium, react very vigorously with the oxygen in the air. They will burn without any heating to form sodium or potassium oxide. These metals have to be safely stored in jars of oil to stop them from coming into contact with the air and reacting! Sodium and potassium are very **reactive** because they both have very unstable arrangements of **electrons**. Other metals, such as magnesium, react easily with oxygen when they are heated in a Bunsen flame. Iron will react similarly—if the iron particles are small enough—to form iron oxide. Metals such as copper and zinc hardly seem to react with oxygen, even if they are heated strongly. They form

A lump of iron will not burn in air, but iron filings (tiny particles of the metal) react and burn quite energetically. We make use of this fact when making fireworks like these sparklers.

a film of oxide on their surfaces in reaction with the air and then do not react anymore. Gold does not react with oxygen no matter how high it is heated, and it is extremely **unreactive** because it has a **stable** arrangement of electrons.

Water games

Some metals react with water by replacing the hydrogen in water. The hydrogen is given off as a gas while the **metal oxide** is left behind. The reactivity of different metals with water varies. Even the smallest piece of sodium or potassium reacts violently with the hydrogen that is given off and catches fire in the heat of the reaction. Calcium fizzes gently in cold water, while magnesium only reacts with hot water or steam. Zinc, iron, and lead show no reaction with water under normal conditions, although iron reacts very slowly (rusts) when exposed to air and water at the same time. Copper and gold do not react at all.

The acid test

When some metals react with **acids**, they take the place of the hydrogen and form a new substance called a **salt**. The hydrogen is given off as bubbles of gas. When sodium and potassium react with an acid, hydrogen is given off so fast it explodes. Calcium and magnesium fizz rapidly, while zinc and iron react much more steadily. Lead reacts slowly with acids, but copper and gold do not react.

SCIENCE ESSENTIALS
Some metals are more reactive than others—they react more easily. Metals can be arranged in order of reactivity, and this is known as the **reactivity series**.

Metal	Reactivity
potassium	Most reactive
sodium	
calcium	
magnesium	
zinc	
iron	
lead	
copper	
gold	Least reactive

zinc in acid

gold in acid

The reactivity series shows the reactivity of different metals. Whatever the reaction, the order of reactivity stays the same.

Titanium

Titanium is the seventh most abundant metal. It is hard, strong, light, and resistant to corrosion, but it is certainly not one of the best-known metals. Titanium is very difficult to extract from its ore by normal methods. Molten titanium attacks furnace linings and absorbs and reacts with gases. In 1947, a process was developed in the United States to extract titanium. Magnesium, which is more reactive, was used to push the titanium out of its ore. Later, sodium was also used. These techniques have allowed the extraction of enough titanium for use in aircraft manufacture and nuclear engineering. However, it is still not widely available.

Pulling Rank

The **reactivity series** helps us to understand the way in which metals react in a variety of different situations. But it's not just metals that are involved.

Push and shove

The reactivity series helps us to predict how easily metals react with other chemicals. For example, chemists can use reactivity to predict what will happen if one metal is added to a solution of another metal **salt**. (A salt is formed when a metal reacts with an **acid**. For example, copper reacts with sulfuric acid to form copper sulfate—which is a salt—and hydrogen.)

A reactive metal reacts enthusiastically with other chemicals and forms very strong bonds with them. Less reactive metals do not react so easily and therefore, they form relatively weak bonds. When a reactive metal is added to a **salt solution**, if it is more reactive than the metal in the salt, it will push that metal out of the solution and take its place in the salt.

SCIENCE ESSENTIALS

A more **reactive element** will push a less reactive element out of a **compound**. This is called a **displacement reaction**. Hydrogen and carbon can be included in the reactivity series.

▲ Just as a child holds on tightly to a parent, so reactive metals hold on to other chemicals. But children do not always want to hold hands—and neither do the less reactive elements!

▲ Iron is more reactive than copper. Therefore, when an iron nail is dipped in a solution of copper sulfate, the iron pushes the copper out of the solution onto the nail and takes its place in the solution.

iron + copper sulfate → copper + iron sulfate

The acid test

When a metal like magnesium or zinc is added to an acid, a reaction takes place. The metal replaces the hydrogen in the acid, forming a salt, and the hydrogen gets "pushed out." However, other metals such as copper do not react with acids because they cannot displace the hydrogen. This means that hydrogen is less reactive than some metals but more reactive than others. It has a place in the reactivity series that is below zinc and above copper.

Carbon is another nonmetal that is included in the reactivity series between magnesium and zinc. The relative reactivity of carbon makes it very useful for extracting metals from their ores.

Metals from ores

The earth's crust is made up of many different types of rock, and many of them contain metals. These metals are rarely pure. They are usually present as chemical compounds—known as metal ores—making up the rock. Getting the metal out of the ore can be tricky, and the way it is extracted depends in part on the position of the metal in the reactivity series. If the metal is lower than carbon in the series, then carbon is often used to displace the metal and free it from its ore.

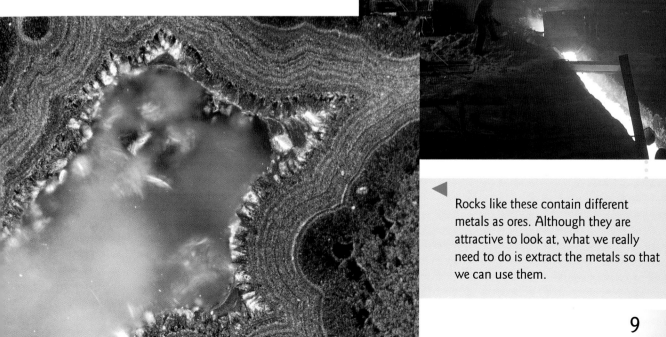

To get iron from lumps of iron ore (like this piece of polished hematite), we need heat and carbon to displace the metal from the rock.

Rocks like these contain different metals as ores. Although they are attractive to look at, what we really need to do is extract the metals so that we can use them.

Using Reactivity

Knowing how different **elements** react helps us to understand what happens to them during chemical reactions and shows us how we can use them more effectively.

> **SCIENCE ESSENTIALS**
> Most metals are found locked up in **compounds** contained in rocks known as ores.
> The way a metal is extracted from its ore depends on its reactivity.
> Metals react with oxygen to form **metal oxides**.

Using heat

Ever since our early ancestors in the Bronze Age discovered the usefulness of metals such as copper, people have extracted as much as possible from the earth's rocks. The less **reactive** metals do not hold on to the other elements in the metal ore too tightly, so they can be extracted quite easily. Simply heating mercury is enough to extract it from its ore. The ore is mainly mercury oxide that splits upon heating into mercury and oxygen in a process known as **thermal decomposition**.

Competitive metals

Many metals cannot be extracted from their ores by simple heating. Some need heating combined with a bit of chemical competition using an element from higher up in the **reactivity series**. For example, carbon is often used to extract metals such as lead and iron from their ores.

Aluminum is the most abundant metal found in the earth's crust, but it is fairly reactive. Because of this, the aluminum oxide in the most common ores has to be split using powerful electric currents in a process know as **electrolysis**. For electrolysis to take place, the aluminum oxide has to be molten, but it will only melt at extremely high temperatures that are expensive to maintain. As a result, aluminum was extremely expensive and its uses were very limited for many years. Eventually, a different method for extracting this light, useful metal was developed. The crushed ore was mixed with the mineral cryolite to create a mixture that melts at a relatively low temperature and is ready for electrolysis. This method made aluminum cheaper to produce and now the metal is widely used in everything from soft drink cans to nuclear reactors.

Rusting away

Iron and steel are some of the most commonly used **materials** around. The only major problem with them is that they rust (react with oxygen and water in the air to form flaky, brown iron oxide). This process eats away at the metal structure and destroys it. The rust then falls away, exposing fresh metal to reaction and decay.

Built-in obsolescence! Sooner or later, every car will be reduced to a heap of flaky, brown rust.

Galvanized into action

The problem of rusting can be solved more or less successfully in a number of ways. The iron or steel can be covered with paint, but this is only effective for as long as the paint stays intact. If the steel is coated in a thin layer of a more reactive metal, such as zinc, the steel will not rust. Zinc reacts with oxygen to form an oxide layer but unlike rust, this oxide layer stays in place and protects the metal underneath from further reaction. Even if the metal gets scratched, any reaction that takes place is between the more reactive zinc, oxygen, and water, leaving the iron unscathed. Metal covered with zinc for this reason is described as galvanized.

Sea water makes iron and steel rust very quickly. To counteract this problem, large blocks of zinc are often bolted onto the steel hulls of ships. The zinc blocks corrode and are "sacrificed" to protect the ship. They need to be replaced regularly.

Oxidation and Reduction

Oxidation and **reduction** are some of the most common reactions in the world around us. They include burning, respiration, and rusting. What goes on in these important reactions?

Oxidation, reduction, and redox reactions

When a chemical reacts with oxygen, it is said to be oxidized. As a forest fire rages, the chemicals in the wood become oxidized. And when fireworks are set off, the chemicals they contain undergo oxidation reactions. The slow appearance of rust on a car is also an example of oxidation. It is called oxidation when a **compound** gains oxygen or loses hydrogen. The events of a chemical reaction can be shown as simple word equations like these:

> metal + oxygen → metal oxide
> magnesium + oxygen → magnesium oxide

Reduction is the removal of oxygen from a compound or the addition of hydrogen. Very often, the oxidation of one substance is linked to the reduction of another. When the two are linked in this way, it is known as a **redox reaction**. For example:

> iron oxide + carbon → iron + carbon dioxide
> $2FeO + C \rightarrow 2Fe + CO_2$

The iron oxide is reduced to iron (loses oxygen) and the carbon is oxidized to carbon dioxide (oxygen added). Redox reactions like this are valuable in the extraction of metals from their ores.

Safety first

Powerful oxidizing agents can be very dangerous because they provide oxygen, which allows **materials** to burn more fiercely. This means that oxidizing agents are very dangerous near fires. Anything containing an oxidizing agent—such as a tanker truck on the road, a jar in the laboratory, or a bottle of bleach in the home—must carry a hazard symbol to alert everyone to the possible danger.

▶ Tanker trucks on the road often carry chemicals which, like this oxidizing agent, can be dangerous. Warning symbols like this one help make other road users and emergency services aware of the problems they are dealing with if there is an accident.

The living world of redox

Redox reactions are of enormous importance in living organisms.

All living things make or eat the food they need to supply them with energy. This food energy is made available to each animal and plant through a series of carefully controlled redox reactions that take place within every living cell. The biological **catalysts** or **enzymes** that bring about these redox reactions are found on membranes inside tiny rod-like structures known as **mitochondria**. These miniature powerhouses are found in every living cell.

SCIENCE ESSENTIALS

Oxidation is either the combination of chemicals with oxygen or the removal of hydrogen atoms.
Reduction is the removal of oxygen, or the addition of hydrogen, to a chemical.

▼ The power of a tiger is just one example of the energy released by redox reactions.

Acids and Alkalis

If your drains are blocked by fatty, greasy food, one way to clear them is to pour hot caustic soda solution down the drain. A lemonade drink mix does not taste right without the addition of some citric **acid**, the chemical that makes lemons taste sour. Caustic soda and citric acid are similar-looking white crystals, yet one is a relatively harmless acid and the other a dangerous **alkali**.

SCIENCE ESSENTIALS
Some natural chemicals have a sour taste. These are called acids. Strong acids are corrosive.
Bases are the opposite of acids, and alkalis are bases that dissolve in water. Strong bases are corrosive.
Indicators can be used to identify acids and alkalis dissolved in water. The strength of acids and alkalis can be measured on the **pH scale**.

Acids

Acids such as citric acid, sulfuric acid, and ethanoic acid are chemicals that can be very corrosive. They can eat away at clothing, metal, and even human flesh. All acids have a very sour taste, although many of them are too dangerous to put in your mouth. They are often used in chemical reactions in the laboratory, yet acetic acid (vinegar) and citric acid (the sour tasting substance in citrus fruit and fizzy drinks) are examples of acids that are regularly eaten.

The acid enemy

The shiny white enamel that covers our teeth is the hardest substance produced by the human body. If we are lucky, our teeth will last us a lifetime, but their single biggest enemy is acid. Without regular brushing, acid can build up and eat its way through the protective enamel, allowing bacteria to penetrate the living tooth and destroy it. Anything we can do to keep the environment of our teeth neutral or slightly alkaline will preserve dental health. Our saliva is naturally slightly alkaline, toothpastes are alkaline, and when dentists fill a cavity, they often include alkaline material before filling the tooth. Even special chewing gum is now available to help keep up the pH of our mouths!

▶ Sugary substances provide food for the millions of bacteria that live in our mouths and their waste products are very acidic. If we don't clean our teeth regularly, this acid builds up and leads to tooth decay.

Bases and alkalis

Bases are the opposite of acids in the way they react. Because alkalis are bases that dissolve in water, they are the most commonly used bases. Alkalis, like acids, are often used for cleaning. Alkalis react with, and remove, grease and fat. Soaps, detergents, and oven cleaners are all alkalis. Soap is made using the same reaction between alkali and fat. Traditionally, sodium hydroxide was boiled with animal fats, but now a wide range of other fats and oils are used in soap making.

Like strong acids, strong alkalis are corrosive and cause great damage to skin by reacting with the fats and oils found in it. Both acids and alkalis only show their properties when they are dissolved in water.

Using indicators

Many acids and alkalis look very similar, and most form colorless **solutions**. You cannot taste them to see if they are sour if you value your tongue, so telling them apart would be very difficult without **indicators**. Indicators are special chemicals that change color with acids and alkalis. Litmus paper is a well-known indicator that turns red in acid and blue in bases. However, there are many more, including some natural ones such as the juices of red cabbage. For example, red cabbage juice is purple-red when neutral, bright red in acid solution, and blue-green in alkali solution.

The pH scale

We use the pH scale to tell us the strength of an acid or alkali. It runs from 0 (the strongest acid) to 14 (the strongest alkali). Universal indicator is a very special indicator that turns different colors in different strengths of acid and alkali. Anything in the middle of the pH scale (pH 7) is neutral—neither acid nor alkali.

▶ The pH scale

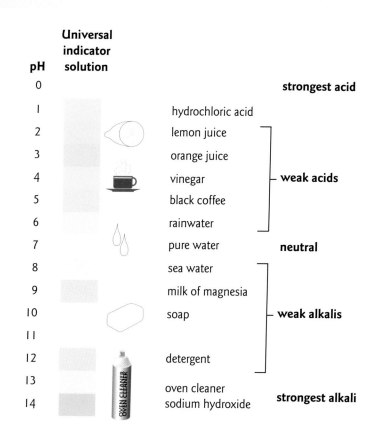

pH	Universal indicator solution		
0			**strongest acid**
1		hydrochloric acid	
2		lemon juice	
3		orange juice	
4		vinegar	weak acids
5		black coffee	
6		rainwater	
7		pure water	**neutral**
8		sea water	
9		milk of magnesia	
10		soap	weak alkalis
11			
12		detergent	
13			
14		oven cleaner sodium hydroxide	**strongest alkali**

Making Salts

Acids and alkalis are useful compounds. The compounds that are formed when they react together are often useful too.

> **SCIENCE ESSENTIALS**
> Acids and bases cancel each other out in **neutralization reactions**. When an acid reacts with a **base**, a **salt** is formed along with water.
>
> acid + base → salt + water

Canceling out

When an acid reacts with a base, the result is a neutral solution. This neutral solution is a compound known as a salt dissolved in water.

acid + alkali → salt + water

nitric acid + potassium hydroxide → potassium nitrate + water

The salts made when acids and bases react have a wide variety of uses. For example, potassium nitrate is used in explosives and fireworks, calcium sulfate is used in plaster, iron sulfate is used to treat anemia, and copper chloride helps cure fungal infections in pet fish!

Gut feelings

After we swallow food, it passes down into our stomach, which produces special **catalysts** called **enzymes**. These break food down into smaller **molecules** that the body can use. Stomach enzymes work best in an acid pH, so the stomach also produces strong hydrochloric acid. The lining of the stomach is protected against this acid. To protect the rest of the digestive system, the partly digested food leaving the stomach is mixed with a liquid called bile. Bile is collected by the gall bladder and is strongly alkaline. It neutralizes the stomach acid before it can do any harm. However, if any of the stomach contents are squeezed back up the esophagus, there is no such protection. The acid burns your esophagus and you feel the pain of indigestion.

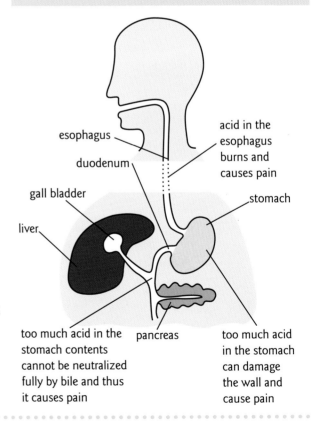

▼ Stomach acid in the wrong place—or just too much of it—can cause painful indigestion.

- esophagus
- duodenum
- gall bladder
- liver
- acid in the esophagus burns and causes pain
- stomach
- too much acid in the stomach contents cannot be neutralized fully by bile and thus it causes pain
- pancreas
- too much acid in the stomach can damage the wall and cause pain

Acid soil

Not only people suffer from indigestion. Sometimes the land itself becomes too acidic. Most plants grow best in neutral or very slightly acidic soil conditions but sometimes, particularly when the soil is peaty, the pH can drop to as low as 3 or 4. To restore some balance to the soil, farmers spread it with lime (an alkali). This reacts with the excess acid to neutralize it and allows the plants to grow well again.

Vinegar for wasps, bicarbonate for bees . . .

Many different animals and plants have evolved stings as a defense against other animals that want to eat them. Nettle leaves are covered with hairs that are like tiny hollow hypodermic needles. If you touch a leaf, an irritant acid solution is injected into your skin that produces a painful rash. Similarly, bees and wasps produce a poison and, when threatened or angry, they will use it to sting. A big difference between these insects is that a wasp sting is an irritant alkaline compound while the bee sting, like the nettle's, is acidic. To reduce the effect of the stings, you can neutralize them as quickly as possible. If you do not have a commercial spray or cream handy, putting vinegar (acid) on a wasp sting and baking soda (alkali) on a bee sting will neutralize the poison and ease the pain.

▶ This close-up of a nettle leaf shows the stinging hairs all too clearly!

Rain and Rocks

The drops of rain that fall on Earth have a surprisingly large effect on the rocks that make up the surface of our planet.

> **SCIENCE ESSENTIALS**
>
> Rocks are broken down by both **physical** and **chemical processes**. These include weathering as a result of the freezing of water. Burning **fossil fuels** is the main cause of **acid rain** that attacks rocks and buildings and kills plants and animals.

Making soil

Water has always played a major part in the breakdown of the rocks from which soil is made. Rainwater and dew fill up small cracks in the rocks. When the water freezes, it expands and pushes the rocks around the crack further apart. This process is repeated many times until loose bits of rock fall away. These are often washed away in streams. Driven along by the water, they bang against each other and grind each other down into smaller and smaller bits. Some particles also react chemically with the water until eventually sand or soil is formed.

This process of the breaking down of rocks is known as weathering. As far as we know, weathering has been vitally important in the formation of soil—ever since the surface of the newly formed Earth cooled and solidified. More recently, the interactions between rain and rocks have taken on a new and more sinister form, and human beings are responsible.

Acid rain

When rain falls, it dissolves a tiny amount of the carbon dioxide in the air, making it a very weak solution of **acid**. With the increase in the burning of fossil fuels (coal, oil, and natural gas) over the last century, more acidic gases have built up in the atmosphere. Levels of carbon dioxide, sulfur dioxide, and nitrogen oxides have increased greatly. These gases also react with rain to form acidic solutions. For example:

> sulfur dioxide + oxygen + water → sulfuric acid

Rain with a pH as low as 2.4 has been measured in some places. In the heavily industrialized northeastern states, women have reported that while walking along city streets, acidic raindrops have eaten holes in their tights and stockings. Acid rain like this has devastating effects on an environment. It may not only affect the area around the source of the pollution—damage may be caused far away, even in another country.

A. The number of vehicles is increasing daily. Each vehicle emits acidic gases from its exhaust. Even those with catalytic converters lose CO_2.

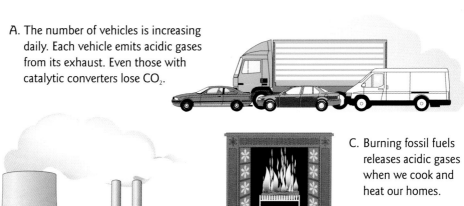

B. Factories and power stations burn oil, gas, and coal to produce electricity, and they release acidic gases.

C. Burning fossil fuels releases acidic gases when we cook and heat our homes.

D. The acidic gases may be blown hundreds of miles in the atmosphere, even to another country. Eventually, they dissolve in water droplets in clouds and fall as acid rain.

E. Where acid rain falls, it directly destroys living organisms such as these trees. Huge areas of forest have been lost in this way. It also destroys life indirectly by making lakes and ponds too acidic for water life, including fish, to survive.

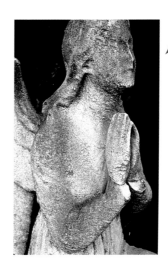

F. Acid rain reacts with limestone rocks and with limestone and marble buildings and statues. The stones dissolve away, spoiling the statues and even making buildings unsafe.

What can be done?

The issue of acid rain has to be addressed by the global community, because acid emissions from one country often fall as acid rain in another. Using energy sparingly, heating our homes a little less, and cutting back on the use of cars would all reduce acid rain. Scientists, engineers, and industrialists have already made great progress in the development of cleaner, almost pollution-free factories. Some ponds and lakes have been "brought back to life" with the addition of lime to neutralize the acidic water. However, to prevent the current levels of acid rain from rising and to reduce future damage, more work needs to be done.

Water

Almost two-thirds of the surface of Earth is covered in water. The properties of this very special liquid have made life as we know it possible.

> **SCIENCE ESSENTIALS**
> Substances such as water expand and contract with changes of temperature. Water is constantly being recycled through the water cycle.

The properties of water

Water is the common name for the chemical H_2O—two **atoms** of hydrogen are joined to one atom of oxygen in a particular way. Water is vitally important to living things for a number of reasons.

Water is an unusually good solvent, and an enormous range of very different chemicals will dissolve in it. This makes it possible for the seas and oceans to be made of **salt** dissolved in water. At the same time, large carbon-containing **molecules** are dissolved in the water that makes up around 70 percent of all cells. Water's amazing properties as a solvent make all the reactions of life possible.

The density of most substances changes as they heat up or cool down, but the density of water as it cools changes in an unusual and very important way. As water cools to below 39°F (4°C), the molecules arrange themselves differently and take up more space. As the temperature drops to 32°F (0°C) and the water freezes, the ice that forms is less dense than the liquid water, and so it floats. This is a unique property of water. It is also very useful. For example, the fact that ice floats on water means that living things can survive in ponds and rivers even when temperatures fall below freezing.

Ice acts as an insulating layer, which helps prevent the rest of the water from freezing. If ice formed from the bottom up, life would only be found in ponds, rivers, and oceans in regions where the water never freezes.

Rapid changes in temperature cause changes in the rates of chemical reactions, so they are bad news for living organisms. Water is very slow both to absorb and release heat, which means lakes, oceans, and seas do not heat up or cool down very much as the weather changes. This stability in a lake or ocean environment is good news for the animals and plants that live in the water and helps to explain how life might have evolved originally in a water-based mixture of chemicals.

The water cycle

The earth's water is being recycled constantly through animals, plants, and the atmosphere. The water you drink has already passed through the kidneys of many thousands of other living organisms!

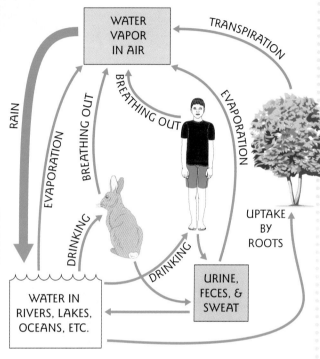

▶ How the water cycle moves through living things

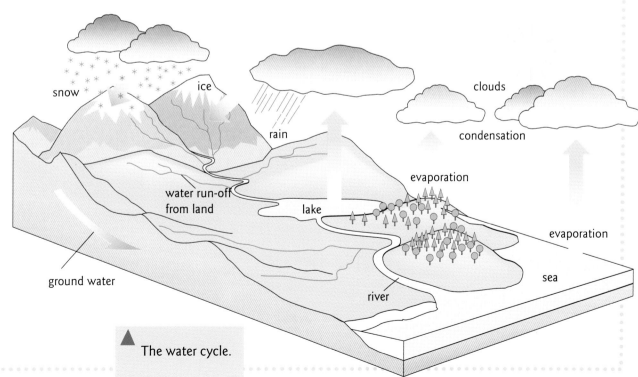

▲ The water cycle.

21

Pollution

Water and air are our most precious resources. Without them, human life—all life on the earth—would perish. Yet increasingly, we are polluting both the air and water with our waste and the chemicals we use in our ever-developing way of life.

> **SCIENCE ESSENTIALS**
> Some of the technological progress made by humans has damaging effects on the environment.
> Burning **fossil fuels** has many implications for the environment.
> Toxic material can accumulate in food chains and water courses.

Polluting the water

As the human population soars, we release more and more of our waste products into the rivers and seas. Most of this sewage is untreated, which creates enormous risks of disease and **eutrophication** due to increased nitrate levels. The widespread use of nitrate fertilizers to increase crop yields adds to this problem. We also pollute water with pesticides, oil spills from shipping, and many waste products from factories and chemical plants. The effects of such pollution on organisms are sometimes dramatic, causing immediate death or serious damage. But it may be less obvious, causing small cell faults that could lead to diseases such as cancer.

1. When a source of nitrates (fertilizers, manure, or human sewage) enters a river, it is used as a source of nutrients by the water plants, which then grow profusely.

2. When the plants die, microbes break them down and use up large amounts of oxygen from the water.

3. As a result, fish and other water animals die and decay, using up more oxygen and releasing more nitrates.

4. Eventually the water, choked with weeds, can no longer support any animal life. This is eutrophication.

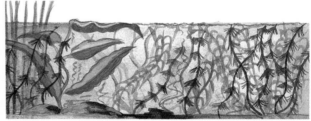

Eutrophication—the lingering death of a river.

Air pollution

We constantly breathe air in and out of our lungs—so any air pollution is taken right into our bodies. As the human population has increased, so has the use of fossil fuels. As fossil fuels are burned, they produce the gases that result in **acid rain**. Fossil fuels also produce **greenhouse gases** such as carbon dioxide and carbon monoxide. As vast areas of rainforests are cut down and burned, pollution is released into the air, and the trees that use up carbon dioxide are lost. The large numbers of cattle raised to produce meat, such as hamburgers, pass huge amounts of methane (a greenhouse gas) out of their digestive systems. The world climate appears to be changing and getting warmer. Many scientists think this is due to pollution and the buildup of greenhouse gases.

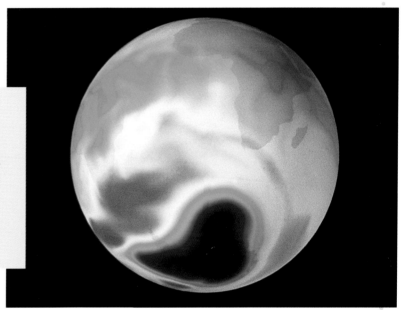

▶ CFCs (**chlorofluorocarbons**) damage the ozone layer and destroy our protection against harmful ultraviolet radiation. This affects our climate. The dark blue area on this image is a hole in the ozone layer over the Antarctic.

What can be done?

Because human activity is the main cause of pollution, the simple answer would be for us to stop doing the things that cause pollution. But that is easier said than done. Almost everyone would agree that global warming is a bad thing and carbon dioxide emissions should be controlled—but hardly anyone wants to stop driving their car.

The **developed world** may not want the **developing world** to produce more pollution, but of course, those in the developing world want to progress and enjoy an improved standard of living too. Gradually, by negotiation and hard work, progress is being made, and it is in all of our interests for it to continue.

Materials

Everything around us that has been constructed by people is made of **materials**. Some of these materials are naturally occurring, and others are synthetic.

> **SCIENCE ESSENTIALS**
> Different materials have different properties that make them useful for different tasks.
> Synthetic materials are developed by making new **compounds** through carefully controlled reactions.

Natural materials

Since the earliest recorded stages of human evolution, people have used natural materials to make the things they want and need. Horn and bone were strong and smooth enough to be used for tools, weapons, and jewelry. Animal hides were found to be flexible and warm so they were used for warm clothing, seats, and bedding along with woven plant material. Clay is soft when "raw," but upon heating, it becomes hard and brittle and an ideal material for holding food or even water.

The first buildings were made of wood, stones, and mud. When metal was discovered, money and more sophisticated weapons and jewelery were made. As time passed, people demanded more of the materials they used, and over the last century, more and more new **synthetic substances** have emerged.

▲ Stones such as flints are hard and can be given sharp edges. At first, humans were relatively weak animals who were easy prey, but with the use of stone weapons, they developed into creatures to be reckoned with, even by top predators such as saber-toothed tigers.

Synthetic materials

Many new materials have been developed from the products of the oil industry. The larger organic (carbon-containing) **molecules** made when **crude oil** is distilled to produce fuels can be split into smaller molecules in a process known as "cracking." These molecules can then be joined together again in different ways. Often, this involves the formation of long chain molecules known as **polymers**. These plastics have revolutionized everything from food packaging to engine design. Plastics do not corrode, they are often very tough, and most importantly, they are usually cheap.

Nylon

Nylon was developed in 1934 by Wallace Hume Carothers, who was working for the Du Pont Company. He invented a revolutionary new fiber called nylon that went on the market in 1935.

Nylon is used in stockings and tights, but it has many other uses too. Added to other fibers such as wool in carpets, it adds durability. Nylon ropes do not rot; nylon bearings do not wear. The list of uses of this synthetic material is almost endless. Sadly, Carothers did not live to see the success of his great development. Poor, relatively unrecognized, and jilted in his personal life, he killed himself before nylon went on the market. Who knows what else he might have discovered?

Rot-ability?

Natural materials rot as they are oxidized or slowly broken down by the action of **microbes**. The problem with many synthetic materials is that the microbes do not exist to break them down. This means that waste plastics and waste nylon do not disappear. As awareness of the problem of plastics grows, companies are beginning to use **biodegradable plastics** that can, sooner or later, be broken down by microbes.

▶ The vast amounts of plastic dumped in the environment will always be with us.

New materials for old

In an attempt to improve on nature, some synthetic materials have been made in imitation of natural ones. Instead of natural rubber, synthetic rubber can be made with the properties of grip and wear that are ideal for a particular job. While synthetic diamonds are not as rare and beautiful as natural ones, they are a great deal cheaper when you need a drill bit instead of a necklace!

Fossil Fuels

For hundreds of years, the **developed world** has increasingly depended upon the use of **fossil fuels**. During the last hundred years, the role of fossil fuels in daily life has become even more important. Today, it is impossible to imagine life without them, yet our supply is limited and one day we will run out.

> **SCIENCE ESSENTIALS**
>
> Fuels contain stored energy that is given off when they burn in air (combustion). Fossil fuels include coal, oil, and gas. These contain energy from the sun, trapped when they existed as animals and plants millions of years ago.

Fossilized sunshine

Millions of years ago, the land was colonized by enormous plants—the distant relations of mosses and ferns we see today. They trapped the energy from the sunlight by **photosynthesis** and used it to make their leaves, roots, and stems. As the plants died and were buried deep in the earth, layer upon layer of the plants became fossilized to form the black substance we know as coal. Now buried deep underground, the coal has to be mined in a variety of ways.

Similarly, in prehistoric oceans, tiny marine plants photosynthesized and were eaten by small animals. The bodies of these tiny animals and plants (plankton) were covered in layers of sand and mud. This stopped them from decaying. Instead, with heat and pressure over millions of years, they broke down to form **crude oil** and **natural gas**.

The usefulness of oil was first discovered in Pennsylvania in 1859. Since then, oil wells have been established all over the world wherever oil is found. Oil and the gas that is usually found with it are of enormous value and many owners of oil-rich land have become very wealthy.

Using fossil fuels

Heating has always been a major use of fossils fuels and it still is today. The three main fossil fuels—coal, oil, and gas—are used to heat our homes and water. They are also used to power the generation of electricity by heating the water that drives steam turbines in power stations. Crude oil in particular has a much wider range of uses. It can be split in the process of **fractional distillation**. Many of the **fractions** are used as fuels for different types of engines, but they are also used to make a wide variety of chemicals including plastics such as **polystyrene**.

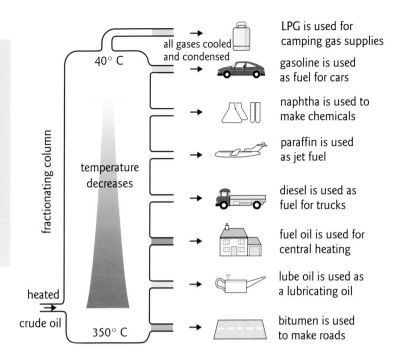

Fractional distillation. The crude oil is heated until it boils and the gases are passed through the fractionating column that is hot at the bottom and cold at the top. The different chemicals that make up the crude oil mixture (known as fractions) are collected at different levels of the tower and are ready to be used in a variety of ways.

The downside . . .

Fossil fuels are immensely useful and their price on the world market is one of the main factors affecting the global economy. There are two major problems associated with our use of these valuable fuels. One is that the supply will not last forever. There is a finite amount of coal, oil, and gas beneath the surface of the earth. Eventually, we will use it and no more will form. It is essential that we develop alternative energy sources before the fossil fuels run out.

A second problem with the use of fossil fuels is that it pollutes the atmosphere. As fossil fuels burn, they produce sulfur dioxide and nitrogen oxides. These cause **acid rain** and smog. They also produce carbon dioxide and monoxide, which are **greenhouse gases** that build up in the atmosphere and cause global warming. As people become increasingly aware of the damage being caused, more effort is going into preventing pollution and protecting our precious environment.

This wind farm in California uses the energy of the wind to drive turbines and generate electricity without polluting the atmosphere. Alternative energy sources are available. We just need to invest in them.

Looking Back

The science of chemistry has emerged in the last couple of centuries from a fascinating tangle of events.

Chemists by accident . . .

Once our early ancestors began to use fire, it was only a matter of time before they unknowingly mastered the chemistry of metal extraction—resulting in what we know as the Bronze Age, which was followed by the Iron Age. The first people known to have deliberately carried out experiments were the ancient Egyptians. Through experiments many thousands of years ago, they managed to produce glass and the blue dye indigo. The term "chemistry" comes from the word used by the ancient Greeks to describe Egypt, the country where so many experiments were carried out.

The Greeks carried on this tradition, although more by discussion than experimentation. The Greek **philosophers** first suggested that **atoms** were the smallest particles of matter. Although they were on the right track, their ideas about the four **elements** (earth, fire, water, and air) have long since been rejected.

. . . and chemists by design

The first work that can truly be labeled as the forerunner of modern chemistry is that of the **alchemists**. Alchemy was a mixture of scientific investigation, mystical quest, and philosophy. Its main goals were to discover the elixir of life that would give whoever drank it eternal life and the philosopher's stone that would turn base metal to gold. The alchemists continued their quests for about 2000 years, and although they never unearthed the fabulous elixir or the famous stone, they did discover strong acids, the distillation of alcohol, the element zinc, and the use of opium as a painkiller.

Even Sir Isaac Newton, who is always remembered for his work on gravity, was a famous alchemist. He published as many books on alchemy as he did on science and mathematics.

▶ An alchemist at work

The dawn of the Scientific Age

Toward the end of the 18th century, there was a move away from alchemy and towards chemistry as we know it today. Antoine and Marie Lavoisier in France and Joseph Priestly and John Dalton in England began working and writing with true scientific method on the chemistry of air and the atomic nature of matter. More accurate techniques of observation and measurement were used, and detailed records of experiments were kept so that others could attempt to repeat them.

> ▶ France's big mistake? Antoine Lavoisier was one of the greatest minds ever produced by France. With his wife Marie, he was one of the first true chemists whose ideas and experiments were well ahead of his time. In spite of this, he was beheaded on false charges by the fanatics of the French Revolution. His death was a loss to the whole world.

SCIENCE ESSENTIALS
Chemistry as we recognize it today emerged from the alchemists of earlier centuries.
Our understanding of chemistry and of chemical **models** has changed over the years and is still changing today. Many of the early theories, however, are now accepted as facts.

Forward to the future

Over the past 200 years, chemists have solved, at least partially, many of the great questions being asked in their science. However, new and exciting discoveries continue to be made. Carbon has long been known and recognized as one of the most common elements and is found in one of three forms (soot, graphite, or diamond).
Yet in the late 20th century, the British scientist Harry Kroto found an entirely new form of the element. It is called **buckminsterfullerene** (or buckyballs for short!) because its shape resembles the **geodesic domes** of American inventor R. Buckminster Fuller. The "new" material has superconducting properties and it has opened up whole new areas of chemistry.

Most of the chemistry we have discovered has benefited the human race and improved not only our understanding of matter but also our quality of life. Yet some has undoubtedly caused damage, both in the form of pollution and in deliberate acts of warfare. Who knows what we will discover in the future—and how we will use the knowledge we gain!

Glossary

acid chemical with a sour taste that is often corrosive and has a low pH factor

acid rain rain in which acidic gases have dissolved, lowering the pH factor

alchemist forerunner of modern chemists who sought the elixir of life and the philosopher's stone

alkali base that dissolves in water

atom smallest particle of an element

base chemical with a bitter taste that has a high pH factor; the opposite of an acid

biodegradable plastics plastics that can be broken down by microbes

buckminsterfullerene new form of the element carbon

catalyst substance that speeds up the rate of a reaction

chemical equation form of mathematical equation used to notate chemical reactions

chemical process involves chemical reactions

chlorofluorocarbons (CFCs) chemicals that damage the ozone layer

compound substance made of two or more different types of atoms chemically joined together

crude oil fuel made from the fossilized remains of living things millions of years ago

developed world those countries such as the United States and much of Europe that have exploited their resources in order to gain a relatively high standard of living for their populations

developing world those countries, especially in much of Africa and South America, which are currently preparing their resources for greater exploitation in order to improve their populations' standard of living

displacement reaction reaction in which a more reactive element pushes a less reactive element out of a compound

electrolysis splitting a compound using electricity

electron negative particle that travels rapidly around the nucleus of an atom

element substance made of only one type of atom, all having the same number of protons in their nuclei. Over 100 elements have been discovered, but only 91 of them occur naturally.

enzyme biological catalyst that speeds up the rate of reactions in living cells

equation *see* **chemical equation**

eutrophication excess growth of plants following nitrate pollution in a waterway. It results in a severe lack of oxygen in the water as microbes bring about the decay of the dead plants.

food chain links between different animals that feed on each other and on plants

formula tells us the type of atoms combined in a compound and the numbers of them in each molecule

fossil fuels coal, oil, and natural gas fuels made from the fossilized remains of living things millions of years ago

fractional distillation separating a mixture of liquids by distilling each one at a different temperature

fractions mixture of different chemicals that make up the crude oil mixture

geodesic dome domed or vaulted structure of straight elements that form interlocking polygons

greenhouse gases gases such as carbon dioxide and methane that increase the greenhouse effect and bring about global warming

indicator substance used to identify acids and alkalis in solution in water

materials stuff out of which everything made by people is constructed

metal oxide compound formed when a metal reacts with oxygen

microbes very small organisms, such as bacteria and protozoa, that can only be seen clearly with a microscope

mitochondria tiny organelles (compartments) found in all living cells. They are the site of respiration and source of energy for the cell.

model idea used to help understand the way things work

molecule particle made up of more than one atom joined together

natural gas type of fossil fuel

neutralization reaction reaction between an acid and a base, which results in the formation of a neutral solution of a salt in water

oxidation type of reaction that commonly involves either the combination of chemicals with oxygen or the removal of hydrogen atoms

pH scale used to measure the strengths of acids and bases

philosopher someone who inquires into the nature of things using logical reasoning rather than scientific experiments

photosynthesis process by which green plants make food from carbon dioxide and water using the sun's energy

physical process may involve a change in state but does not involve chemical reactions

polymer long chain of molecules made up of many small repeating units

polystyrene plastic often used for packaging and take-out food containers

products chemicals produced as the results of a reaction

reactants chemicals that react together in a given reaction

reactive substance (or atom) that has an unstable arrangement of electrons

reactivity series arrangement of metals and other elements in order of their reactivity

redox reaction reaction where the oxidation of one substance is linked to the reduction of another

reduction removal of oxygen or the addition of hydrogen to a chemical; the opposite of oxidation

salt when a metal and an acid react together, a salt is formed along with hydrogen gas

salt solution a salt dissolved in water

stable describes an atom with an arrangement of electrons that makes it unlikely to lose or gain other electrons

synthetic substance material made by people rather than naturally occurring

thermal decomposition splitting of a compound by heating

unreactive describes an element or atom with a stable arrangement of electrons that tends not to react with others

More Books to Read

Barber, Jacqueline. *Chemical Reactions*. Berkeley, Calif.: University of California, Berkeley, Lawrence Hall of Science, 1998.

Cunningham, A. *Essential Chemistry*. Tulsa, Okla.: EDC Publishing, 1992.

Mebane, Robert C., and Thomas R. Rybolt. *Adventures with Atoms & Molecules*. Springfield, N.J.: Enslow Publishers, 1998.

Index

acetic acid 14
acid rain 18–19, 23, 27
acids 7, 8, 9, 14, 15, 16, 17, 28
air pollution 22, 23
alchemy 28, 29
alkalis 14, 15, 16, 17
alternative energy sources 27
aluminum 10
atoms 4, 5, 20, 28
bases 14, 15, 16
bile 16
buckminsterfullerene 29
calcium 7
carbon 8, 9, 10, 12, 19, 29
carbon dioxide 12, 18, 19, 23, 27
carbon monoxide 23, 27
catalysts 13, 16
caustic soda 14
CFCs (chlorofluorocarbons) 23
chemical equations 5
chemical models 4, 29
chemical processes 18
chemical reactions 4, 5, 10, 14, 21
chemistry 28, 29
citric acid 14
coal 18, 19, 26, 27
compounds 4, 5, 8, 9, 10, 12, 16, 24
copper 6, 7, 8, 9, 10
corrosion 7, 14, 15
cryolite 10
density 20
developed world 23, 26
developing world 23, 25
displacement reaction 8
Earth's crust 9, 10
electricity 19, 26, 27
electrolysis 10
electrons 4, 5, 6
elements 4, 5, 10, 29
    reactive elements 5, 8
    unreactive elements 5
energy 13, 26
enzymes 13, 16
ethanoic acid 14
eutrophication 22
formula 4
fossil fuels 18, 19, 22, 23, 26–7
fractional distillation 26, 27
fractions 26, 27
galvanized metal 11
gas 18, 19, 26
global warming 23, 27
gold 6, 7
greenhouse gases 23, 27
hydrogen 4, 6, 7, 8, 9, 12, 13, 20
ice 20
indicators 14, 15
iron 5, 6, 7, 8, 9, 10, 11, 12
iron sulfide 5
Lavoisier, Antoine 29
lead 6, 7, 10
lime 17, 19
litmus paper 15
magnesium 6, 7, 9
mass 5
materials 11, 12, 24–5
    natural materials 24, 25
    synthetic materials 24–5
mercury 10
metal oxides 6, 10, 11, 12
metals 5, 6–11, 12
microbes 25
mitochondria 13
molecules 4, 5, 16, 20, 24
neutralization 16, 17, 19
Newton, Sir Isaac 28
nitrates 22
nylon 25
oil 18, 19, 24, 26, 27
ores 9, 10, 12
oxidation 12, 13
oxidizing agents 12
oxygen 4, 6, 10, 11, 12, 13, 20
ozone layer 23
pH 14, 15, 16, 17, 18
photosynthesis 26
physical processes 18
plants 17, 22, 26
plastics 24, 25, 26
    biodegradable plastics 25
pollution 18–19, 22–3, 27
polymers 24
polystyrene 26
potassium 6, 7
products 5
rainwater 15, 18
reactants 5
reactivity 6–9, 10
    displacement reaction 8
    reactivity series 7, 8, 9, 10
redox reactions 12, 13
reduction 12, 13
rocks 9, 10, 18
rust 11, 12
salt solutions 8
salts 7, 8, 9, 16, 20
sodium 6, 7
soil 17, 18
solutions 15, 16
solvents 20
steel 11
stings 17
stomach acid 16
sulfur 5
sulfuric acid 8, 14
thermal decomposition 10
titanium 7
tooth decay 14
tooth enamel 14
water 4, 6, 11, 15, 18, 20–1, 22
water cycle 20, 21
weathering 18
wind farms 27
zinc 6, 7, 9, 11, 28

32